Guidance To The Commerce Department's

REEXPORT CONTROLS

U.S. DEPARTMENT OF COMMERCE
Bureau of Industry and Security
Office of Exporter Services

OVERVIEW

The United States Department of Commerce regulates exports and reexports of "dual-use" items, i.e., goods, software and technologies with commercial and proliferation/military applications, through its Export Administration Regulations (EAR). If you are outside the United States and wish to export or reexport an item that is of U.S. origin or that has a U.S. connection, your product may require a license from the U.S. Department of Commerce's Bureau of Industry and Security (BIS).

This overview is designed to give an understanding of our regulations and how to use them. However, nothing provided here can substitute for consulting the EAR. The EAR include answers to frequently asked questions, detailed step-by-step instructions for determining if a transaction is subject to the regulations and how to apply for a license. In using the EAR, you may want to first look at Part 732 for the steps you follow to determine your obligations.

If you would like to review a specific part of the EAR referenced in this document, go to the on-line EAR database on the Government Printing Office Web site at ***www.gpo.gov/bis***.

CONTENTS

What is a Reexport?

A ***reexport*** is the shipment or transmission of an item subject to the EAR from one foreign country (i.e., a country other than the United States) to another foreign country. A reexport also occurs when there is “release” of technology or software (source code) subject to the EAR in one foreign country to a national of another foreign country.

Determining whether your item is subject to the Export Administration Regulations.

Your item is subject to the Export Administration Regulations (EAR) if it:

- Was produced or originated in the United States (See Part A);
- Is a foreign made product that contains more than a specified percentage of U.S. controlled content (See Part B);
- Is a foreign made product based on certain U.S.-origin technology or software and is intended for shipment to specified destinations (See Part C); *or*
- Was made by a plant or major component of a plant located outside the United States, and if that plant or major component of a plant is the direct product of certain U.S. technology or software, and your product is intended for shipment to specified destinations (See Part D).

A. Determining whether a U.S.-origin item requires a license from BIS.

You may need to obtain a license to "reexport" an item that was produced or originated in the United States. Many items subject to the EAR do not need a license to be reexported from one foreign country to another. But certain items are controlled and will either require a license or must qualify for a License Exception. License requirements apply particularly to items controlled by multilateral export control regimes. In addition, some destinations and persons (individuals or groups) are subject to comprehensive export controls, including controls on widely traded consumer products.

To determine whether your U.S.-origin product requires a license you will need the following three pieces of information:

1. The Export Control Classification Number (ECCN). Certain items, notably those controlled by multilateral export control regimes, are on the Commerce Control List (CCL) (part 774 of the EAR) and are included in a specific ECCN. If your item is not on the CCL, it may be classified as EAR99. EAR99 is a general category of goods and technology that encompasses many widely traded consumer and industrial items. The ECCN in the CCL will also tell you the reason(s) for control.
2. The ultimate destination of the item. You will need to match the reason(s) for control listed in the ECCN entry on the CCL (part 774 of the EAR) with the country of ultimate destination in the Country Chart (part 738 of the EAR). The reason(s) for control, when used in conjunction with the Country Chart, will help you to determine if a license is required to the ultimate destination. If you determine that your reexport transaction requires a license, you should review the EAR to determine if any License Exceptions are available (part 740 of the EAR). For general information on License Exceptions, see part E below.
3. The end-user and end-use for the item. Even if you determine a license is not required based on the ultimate destination (or a license is required but a License Exception would generally apply), you may need to apply for a license because of the end-use or end-user. There are certain special restrictions that apply to persons (or entities) identified in the EAR, as well as to persons whom you know or have reason to know are involved in weapons proliferation activities. In most instances, a license is required to persons identified in part 744 of the EAR for the reexport of all items subject to the EAR (i.e., all items on the CCL and all items classified as EAR99).

B. Determining whether your foreign-produced product requires a license from BIS because it contains some U.S.-origin content.

As noted above, certain foreign-produced items are also subject to the EAR because they contain more than a specified percentage value of U.S.-origin controlled content. You need to first determine if your foreign produced item is subject to the EAR. If you determine your foreign produced item is subject to the EAR, you will then follow the process outlined in Part A above to determine if your foreign-produced item requires a license.

The following steps are provided as general guidance for determining whether a foreign produced commodity that incorporates U.S.-origin parts is subject to the EAR or qualifies for the *de minimis* exception to the EAR. This general guidance does not take into account specific items that are not eligible for *de minimis* treatment. You should consult section 734.4 and Supplement 2 to Part 734 for information on such items and the calculation of U.S. controlled content.

1. General guidance regarding incorporation of U.S. parts or components into foreign-produced items. If you are a foreign company that incorporates U.S.origin parts or components in a foreign commodity you will need to:
    a. classify the U.S.-origin parts or components exported to you according to the classification system set forth in the EAR. The U.S. exporter may be able to assist you in determining the ECCN.
    b. determine if the U.S. parts or components are "controlled content." ("U.S. controlled content" is content that would require a U.S. license if it were to be reexported as separate parts or components to the country of ultimate destination.)
    c. determine if the U.S. "controlled content" is greater that 25% of the value of your finished foreign product. (For designated terrorist supporting countries, you need to determine if the U.S. "controlled content" is greater than 10% of the value of your finished product.) If the U.S. controlled content is 25% or less of the value of your finished product (or 10% or less for applicable countries), you qualify for the *de minimis* exception set forth in section 734.4 of the EAR and your product is not subject to the EAR.
    d. if the controlled content is greater than 25% (or 10% for applicable countries), your product is subject to the EAR.

If your product is subject to the EAR, you need to determine if your item requires a license, either because of the ultimate destination or the end-use or end-user. To do this, follow the steps outlined in Part A 1-3.

2. *Additional guidance regarding foreign software incorporating U.S.-origin software and foreign technology commingled with or drawn from U.S.-origin technology.* If you incorporate U.S.-origin software into your foreign software or if your foreign technology is commingled with or drawn from U.S.-origin technology, you would follow a process similar to the one outlined above. That process and a related one-time reporting requirement are set forth in section 734.4 and Supplement 2 to part 734 of the EAR. **NOTE:** The *de minimis* exception can only be used for items that are alike - hardware into hardware or software into software. You cannot use the *de minimis* exception for incorporating software into hardware.

C. Determining if your foreign product requires a license from BIS because it is the direct product of U.S. technology or software.

Direct products of U.S.-origin technology are subject to the EAR only if they are intended for specified destinations, would be subject to national security controls (if they were U.S.-origin items) and the U.S.-origin technology or software on which the foreign product is based required a written assurance from the recipient when it was exported from the United States. See EAR sections 734.3(a)(4) and 736.2(b)(3).

D. Determining if your foreign product requires a license from BIS because it is the direct product of a plant or major component of a plant that was developed based on U.S. technology.

The product of such a foreign plant or major component of a plant is subject to the EAR only if it is intended for specified destinations, would be controlled for national security reasons (if it were a U.S.-origin product) and if the technology on which the plant or major component of a plant is based required a written assurance from the recipient when it was exported from the United States. See section 734.3(a)(5) and 736.2(b)(3).

E. Is your product eligible for a License Exception?

In certain instances, if your reexport transaction requires a license, you may be able to use one of the License Exceptions set forth in part 740 of the EAR. A License Exception allows you to reexport an item without applying for a license, provided your transaction meets all the terms and conditions of the License Exception.

Are there any special restrictions I should know about?

You may not reexport an item subject to the EAR to a party whose export privileges have been denied by BIS. Information on parties subject to denial orders is provided on the BIS Web site at ***www.bis.doc.gov***.

Please note that U.S. persons may be subject to additional restrictions under the EAR. See section 744.6 of the EAR. U.S. persons may also be subject to restrictions under other U.S. Government regulations, such as those issued by the Office of Foreign Assets Control of the U.S. Department of the Treasury or other U.S. Government departments or agencies.

Why should you comply with reexport license requirements?

The Department of Commerce has enforcement and protective measures available to it to ensure that recipients of items subject to the EAR comply with the reexport license requirements of the EAR. If the Department of Commerce determines that you have not complied with these requirements and restrictions, it may institute administrative enforcement proceedings, resulting in the possible imposition of civil penalties and/or denial of your eligibility to receive U.S. exports (part 764 of the EAR).

Where to apply for a reexport license?

If your reexport requires a license and is not eligible for a License Exception, you may apply for a reexport license electronically through the Simplified Network Application Process (SNAP). You may find the basic information on the SNAP program on the BIS Web site at ***www.bis.doc.gov***. If you have not submitted an application electronically before, you must first complete a "PIN" request package.

Additional information and contacting BIS

If you need help regarding a reexport license or product classification, or wish to obtain other information on U.S. export controls, please consult the BIS Web site.

In addition, you may contact the Office of Exporter Services, Bureau of Industry and Security, directly at:

P.O. Box 273
U.S. Department of Commerce
Washington, DC 20044
Phone: (202) 482-4811
Fax: (202) 482-2927 or
email via the BIS Web site at *www.gpo.gov/bis*

All requests for information and assistance must be in English.

Disclaimer

The Bureau of Industry and Security provides this Guidance on Reexports for information purposes only. The Guidance does not provide an official interpretation or translation of U.S. export control law or regulations. This information does not relieve the reader of any duties or obligations regarding knowledge of or compliance with all relevant U.S. export control laws and regulations as they appear in the U.S. Code of Federal Regulations and as modified by notices in the Federal Register.

NOTES:

NOTES:

NOTES:

NOTES:

NOTES:

NOTES:

NOTES:

NOTES:

NOTES:

NOTES:

NOTES:

NOTES:

NOTES:

Made in the USA
Monee, IL
04 November 2026

67374681R00015